LABORATORY MOUSE

PROCEDURAL TECHNIQUES

MANUAL AND DVD

LABORATORY MOUSE

PROCEDURAL TECHNIQUES

MANUAL AND DVD

John J. Bogdanske, BA
Scott Hubbard-Van Stelle, AS, CVT, RLATG
Margaret Rankin Riley, BS
Beth M. Schiffman, BS, CVT, RLATG

CRC Press
Taylor & Francis Group
Boca Raton London New York

CRC Press is an imprint of the
Taylor & Francis Group, an **informa** business

CRC Press
Taylor & Francis Group
6000 Broken Sound Parkway NW, Suite 300
Boca Raton, FL 33487-2742

© 2011 by Wisconsin Alumni Research Foundation Laboratory
CRC Press is an imprint of Taylor & Francis Group, an Informa business

No claim to original U.S. Government works

Printed in the United States of America on acid-free paper
10 9 8 7 6 5 4 3 2 1

International Standard Book Number: 978-1-4398-5042-8 (Paperback)

Library of Congress Cataloging-in-Publication Data

Laboratory mouse procedural techniques : manual and DVD / John J. Bogdanske ... [et al.].
      p. ; cm.
   Includes bibliographical references.
   ISBN 978-1-4398-5042-8 (pbk. : alk. paper)
   1. Mice as laboratory animals--Handbooks, manuals, etc. I. Bogdanske, John J.
   [DNLM: 1. Animals, Laboratory--Handbooks. 2. Mice--Handbooks. 3. Laboratory Techniques and Procedures--Handbooks. QY 39]

SF407.M5L335 2011
616'.027333--dc22   2010039394

Visit the Taylor & Francis Web site at
http://www.taylorandfrancis.com

and the CRC Press Web site at
http://www.crcpress.com

contents

disclaimers...ix
introduction xi

Section I
mouse DVD text and voice-over

disease management 1

1   mouse handling/transfer...3

2   mouse restraint...5

3   oral gavage 7

4   one-handed injection technique...9

5   intraperitoneal (IP) injection (one person).............................11

6   intraperitoneal (IP) injection (one-person towel method) 13

7   subcutaneous (SQ) injection 15

8   subcutaneous (SQ) injection (one-person towel method) 17

9   maxillary bleed 19

10   pedal vein blood draw 21

11   saphenous vein blood draw 23

12   tail vein injection 25

13   retro-orbital bleed (anesthesia required) 27

14   retro-orbital injections (anesthesia required) 29

15   ear notching 31

16   ear tagging 33

Section II
mouse procedural technique handouts

17   intraperitoneal (IP) injection 37

18   subcutaneous (SQ) injection 41

19   maxillary bleed 45

20   oral gavage 49

21   pedal vein blood draw 51

22 saphenous vein blood draw...55

23 retro-orbital blood draw..59

24 retro-orbital injection 63

25 tail vein injection...65

26 ear notching/punching and ear tags 69

appendix a: mouse (Mus musculus)......................................73
    physical characteristics/anatomical features.......................74
    behavior 75
    sexing and breeding 76
    sources...77
    husbandry..78
    diet...78
    handling and restraint 79
    identification 79
    stress management and enrichment..................................80
    recognizing pain and distress..80
    common diseases and prevention 81
        Diseases 81
        Prevention 81
    record keeping..81
    protocols...82
    euthanasia 83

appendix b: normative data for the laboratory mouse..............85
    general information 85
    blood and oxygen..85
    experimental information 86
    hematology..86
    breeding..86

appendix c: blood volume...87

disclaimers

This training DVD is not intended for use, and must not be used, by individuals who lack thorough training and understanding of animal biosafety. Rather, this DVD should be used as a resource or refresher for techniques previously learned through animal biosafety training.

The techniques depicted in this DVD must be performed in compliance with institutionally approved animal protocols, observing all institutional and governmental safety regulations with regard to the use of appropriate safety equipment and procedures. If you lack a thorough understanding of the institutional and governmental rules applicable to the techniques depicted in this DVD, you must not perform these techniques.

The techniques depicted in this DVD must be performed only in an approved animal facility under the guidance of animal care specialists.

The techniques depicted in this DVD take place in a controlled laboratory setting and present a risk of serious bodily injury (including death) through the following exposure hazards: sharp objects; chemical or pharmaceutical agents administered to mice; and biohazards, including but not limited to the bodily fluids and waste of mice. These hazards may be avoided only by complying with a thorough laboratory safety program that complies with all applicable institutional and governmental rules. If your facility lacks such a program or if you do not thoroughly understand all aspects of that program, you must not perform the techniques depicted in this DVD.

introduction

The trainers at the Research Animal Resources Center (RARC) at the University of Wisconsin–Madison want to thank you for taking the time to view this training DVD. This DVD should be viewed as a resource or a refresher for techniques previously learned in a biomethodology class or in your own research setting. The procedures and techniques demonstrated are those that are approved by the University of Wisconsin Institutional Animal Care and Use Committee (IACUC) and are commonly used on research mice. If, after viewing this DVD, you find that a specific method or technique is not covered in enough detail or that you would prefer further instruction, please feel free to contact a trainer via e-mail at trainer@rarc.wisc.edu, and we will be happy to assist you. It is our hope that you find this information both educational and useful.

I

mouse DVD text and voice-over

disease management

Surveillance, diagnosis, treatment, and control of disease are integral components of adequate veterinary care.[1] Subclinical microbial, particularly viral, infections can occur in barrier- and conventionally maintained animals. Such infections can seriously compromise experimental protocols by inducing profound changes in immunologic, physiologic, neoplastic, and toxicologic responses in infected animals. Therefore, control and elimination of known pathogens are vital for good science as well as the health and well-being of research animals.

Please note that in the following demonstrations, the technicians are performing the techniques wearing exam gloves and a lab coat. Be sure to check the standard operating procedures (SOPs) of your lab or the requirements of your animal facility to ensure that the proper personal protective equipment is worn while working with animals to prevent possible injury to yourself and the spread of disease.

---

[1] American Association for Laboratory Animal Science. *Laboratory Mouse Handbook*. Memphis, TN: American Association for Laboratory Animal Science, 2006.

mouse handling/transfer

Safely transferring animals from cage to cage or cage to work area is the first task you need to perform effectively when working with rodents. The following demonstration shows you how to transfer mice from a shoebox cage to a tabletop. Watch closely as the technician picks up the mouse by the base of the tail. This is the best method to prevent the possibility of stripping the skin off the tail of the mouse.

Voice-Over: Safely transferring animals from cage to cage or cage to work areas is the first task you need to perform effectively when working with rodents. Watch closely as the technician picks up the mouse by the base of the tail. This is the best method to prevent the possibility of stripping the skin off the tail of the mouse. Be sure to place the mouse back on its feet as soon as possible.

mouse restraint

Proper restraint of the mouse is key to performing techniques such as intraperitoneal (IP) injections, oral gavage, and the maxillary bleed. Although it looks simple when demonstrated, the mouse can be a challenge to restrain.

Pick up the mouse by the base of the tail with your dominant hand. Remove it from the cage and place it on the cage top or a surface that allows the mouse to grab hold. Gently pull back on the tail as this will normally cause the mouse to grab hold of the surface. Position your thumb and index finger over the shoulders of the mouse and directly behind the ears. Slowly and gently force the mouse down onto the surface of the cage top while gathering up the skin with your thumb and index finger. When you have a secure grip on the mouse, lift it off the cage top and turn your hand over, exposing the belly of the mouse. Restrain the tail with either the ring or the pinkie finger. If performed properly, the mouse will not be able to move its head too far side to side. Make sure the chest of the mouse is moving, which is a good indication that the mouse is still breathing.

Voice-Over 1: Hold the mouse tail with your dominant hand, then using your nondominant hand grasp the skin above the shoulders and behind the ears using the tips of your thumb and index finger while applying downward pressure. Be sure to gather up enough of the skin to prevent the mouse from turning its head side to side. Lift the mouse off the surface and be sure to observe its breathing. Mastering this restraint will allow you to safely perform a number of procedures.

Voice-Over 2: Here, the same technique as previously described is shown from the side.

oral gavage

Oral gavage in the mouse is used to administer liquid food, experimental solutions, or medication directly to the stomach. The recommended normal volume is 10 ml/kg.

Restraint is the key to this technique. You must have the ability to properly restrain a mouse to successfully administer the solution. The length of the feeding needle should be the same as the distance from the nose of the mouse to its stomach (just below the rib cage). Once proper restraint is achieved, place the feeding needle in the back of the mouth and tilt the head back until the syringe is parallel with the body of the mouse. The needle will slide down easily when positioned properly. *Do not use force.* (It is possible to perforate the esophagus or force the needle into the trachea.) Proper insertion will be achieved using gravity and the weight of the syringe. If there is any resistance, *stop*, pull the needle out, and try again. Insert the feeding needle until only the needle hub shows. At this time, inject your solution. Slowly remove the needle, following its curve as you pull it out.

recommended supplies

- 1-ml Luer lock syringe
- 22-gauge × 2-inch feeding needle

Voice-Over: Once proper restraint is achieved, determine if the length of the needle is appropriate by measuring the distance from the mouse's nose to its stomach. Place the feeding needle in the back of the mouth and tilt the head back until the syringe is parallel with the mouse's body. The needle will slide down easily when positioned properly. *Do not use force.* Proper insertion will be achieved using gravity and the weight of the syringe. Insert the needle until only the hub is showing; you can now safely inject your solution.

one-handed injection technique

Some injection techniques require the ability to manipulate a syringe using only one hand. The following is a brief demonstration of this technique.

Voice-Over: There are a number of techniques that require the ability to manipulate a syringe using only one hand. Stabilize the syringe with your thumb, index and middle fingers while using the ring finger to work the plunger. You can easily practice this in your lab.

intraperitoneal (IP) injection (one person)

Use the intraperitoneal (IP) method to inject large volumes (20 ml/kg) of an anesthetic agent or medication. Caution should be observed during the holding and injecting process since damage to internal organs can occur if care is not taken to properly restrain the animal and control the insertion of the needle. Performing one-handed injections is necessary for this technique.

Have the needle and syringe ready, loaded with the solution to be injected. (In our demonstration, a 25-gauge needle is placed on a 1-ml syringe.) Restrain the mouse using your nondominant hand. Divide the abdomen into quadrants. Inject into the lower right quadrant of the animal as indicated by the red dot in the illustration. Carefully insert the needle at a 45° angle into the abdomen until you sense the needle "pop" through the skin and abdominal wall. Draw back on the plunger. When you observe a vacuum bubble, then inject your solution. When finished, dispose of the needle and syringe in a sharps container. If while drawing back you get a colored liquid, pull the needle out; discard the needle, syringe, and solution into a sharps container; and start over.

recommended supplies

- 1- or 3-ml syringe
- 25-gauge needle

Voice-Over 1: This illustration shows the mouse abdomen divided into quadrants. It is recommended that IP injections be given into the lower right quadrant as indicated by the red dot. This area, cranial to the knee or dotted line in the picture and the left of the midline, will allow you to avoid accidentally injecting into the testicles of the male mouse if they are retracted into the abdomen.

Voice-Over 2: Insert the needle through the skin and abdominal wall into the lower right quadrant. Draw back on the plunger; if you observe a vacuum, then proceed to inject your solution. If while drawing back you get a colored liquid, pull the needle out; discard the needle, syringe, and solution into a sharps container; and start over.

6

intraperitoneal (IP) injection (one-person towel method)

Use the one-person towel method of intraperitoneal (IP) injection to inject large volumes (20 ml/kg) of an anesthetic agent or medication. Caution should be observed during the holding and injecting process since damage to internal organs can occur if care is not taken to properly restrain the animal and control the insertion of the needle. Performing one-handed injections is necessary for this technique.

Use a sturdy cloth towel opened on a table. Place the mouse at the center of the towel and fold the towel over the mouse with your non-dominant hand, leaving the tail area exposed. Grasp the nearest hind leg and roll the mouse over slightly to expose the abdomen. Insert the needle, angled slightly toward the head, into the lower right quadrant of the animal. You will sense the needle "pop" through the skin and abdominal wall. Draw back on the plunger. When you observe a vacuum bubble, inject your solution. When done, pull the needle and syringe out and dispose in a sharps container. If while drawing back you get a colored liquid, pull the needle out; discard the needle, syringe, and solution into a sharps container; and start over.

recommended supplies

- 1- or 3-ml syringe
- 25-gauge needle
- Towel

Voice-Over: The restraint is the only difference between the standard IP injection technique and the towel method. Keeping the mouse under the towel may be the biggest challenge for this method. Insert the needle into the animal's lower right quadrant. Draw back on the plunger; if you observe a vacuum, proceed to inject your solution. If while drawing back you get a colored liquid, pull the needle out; discard the needle, syringe, and solution into a sharps container; and start over.

subcutaneous (SQ) injection

Subcutaneous (SQ) injections can be administered easily to mice. Restrain by the scruff and insert the needle between the tented folds of skin. Pull back on the plunger of the syringe to verify that a vacuum is created, then inject the solution. In general, no greater than 1 ml should be injected per subcutaneous injection site (25-g adult mouse). Use several sites over the back of the animal if larger volumes must be administered. You can inject up to 10 ml/kg (maximum is 40 ml/kg) of liquid. Performing one-handed injections is necessary for this technique.

recommended supplies

- 1- or 3-ml syringe
- 23-gauge needle or smaller

Voice-Over: While grasping the skin above the shoulders and behind the ears with your thumb and index finger, apply downward pressure to secure the mouse. Insert the needle between the tented folds of skin below your fingers. Draw back on the plunger to verify a vacuum and proper needle placement, then proceed to inject your solution.

subcutaneous (SQ) injection (one-person towel method)

For the one-person towel method of subcutaneous (SQ) injection, use a sturdy cloth towel opened on a table. Place the mouse at the center of the towel and fold the towel over the mouse, leaving the tail area exposed. Restrain the front two-thirds of the mouse with the palm of your nondominant hand while tenting the skin over the hindquarters with your thumb and index finger. Insert the needle at the base of the tented folds of skin. Pull back on the plunger of the syringe to verify that a vacuum is created, then inject the solution. In general, no greater than 1 ml should be injected per subcutaneous injection site (25-g adult mouse). Use several sites over the back of the animal if larger volumes must be administered. You can inject up to 10 ml/kg (maximum is 40 ml/kg) of liquid. Performing one-handed injections is necessary for this technique.

recommended supplies

- 1- or 3-ml syringe
- 23-gauge needle or smaller
- Towel

Voice-Over: Place the mouse at the center of a towel and fold the towel over the mouse, leaving the tail area exposed. Restrain the front two-thirds of the mouse with the palm of your non-dominant hand while tenting the skin over the hindquarters with your thumb and index finger. Insert the needle at the base of the tented folds of skin. Draw back to verify that a vacuum is created, then proceed to inject the solution.

maxillary bleed

The maxillary bleed can be a good alternative to the retro-orbital bleed. Anesthesia is not required with this technique. Depending on the strain of mouse, it is possible to obtain a fairly large blood sample from the maxillary vein. See the Appendix on page 87 for information on calculating the blood volume and maximum blood sample volume of the mouse. The trainers at the Research Animal Resources Center (RARC) recommend using 10% of the total blood volume as the maximum amount that can be collected at one time. To allow for an adequate recovery period, we further recommend that the maximum amount be withdrawn only once every 2 weeks (based on 10%). Daily samples can be collected as long as the cumulative blood volume does not exceed the calculated maximum over a 7-day period.

Restraint is the key to this bleeding technique. In the following clip, notice how the whisker mark on the jawbone serves as a landmark. The vein is located just behind the whisker mark, in the groove past the jawbone. A meaningful poke and the proper restraint are required to initiate and maintain the flow of blood. Approximately 3 drops of blood equals 100 µl. Once the desired amount of blood is collected, release the mouse from your restraint and apply a clean compress to the bleeding site if necessary.

recommended supplies

- 5.5-mm lancet
- Blood collection tube
- Tube rack

Voice-Over 1: This illustration shows the location of the maxillary vein in relation to the whisker mark on the mouse. The yellow dot indicates the general area where the lancet needs to be inserted.

Voice-Over 2: A meaningful poke and the proper restraint are required to initiate and maintain the flow of blood. Approximately 3 drops of blood equals 100 microliters. Once the desired amount of blood is collected, release the mouse from your restraint and apply a clean compress to the bleeding site if necessary.

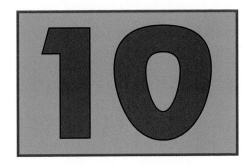

pedal vein blood draw

The pedal vein blood draw technique is used to collect relatively small volumes of blood. See the Appendix on page 87 for information on calculating the blood volume and maximum blood sample volume of the mouse. The trainers at the Research Animal Resources Center (RARC) recommend using 10% of the total blood volume as the maximum amount that can be collected at one time. To allow for an adequate recovery period, we further recommend that the maximum amount be withdrawn only once every 2 weeks (based on 10%). Daily samples can be collected as long as the cumulative blood volume does not exceed the calculated maximum over a 7-day period.

Place a tourniquet as high up on the rear leg of the mouse as possible. Apply a drop of artificial tears (petroleum jelly) near the top of the foot and rub across the surface. The ointment will allow the blood to bead up on the skin. Using a 27-gauge needle, gently poke into the vein, taking care not to pierce completely through it because this will result in blood accumulating under the skin. As the blood beads up, you can use a heparinized capillary tube to collect the blood. When finished, release the tourniquet and apply pressure with gauze to stop the bleeding.

recommended supplies

- Heparinized capillary tube
- Artificial tears (petroleum jelly)
- Tourniquet
- 27-gauge needle
- Gauze

Voice-Over 1: Place a tourniquet as high up on the mouse's rear leg as possible. Apply petroleum jelly near the top of the foot and rub across the surface. The ointment will allow the blood to bead up on the skin.

Voice-Over 2: Using a 27-gauge needle, gently poke into the vein, being careful not to pierce completely through it as this will result in blood accumulating under the skin. As the blood beads up, you can use a heparinized capillary tube to collect the blood.

Voice-Over 3: When finished, release the tourniquet and apply pressure with gauze to stop the bleeding.

saphenous vein blood draw

The saphenous vein blood draw technique is used to collect relatively small volumes of blood. See the Appendix on page 87 for information on calculating the blood volume and maximum blood sample volume of the mouse. The trainers at the Research Animal Resources Center (RARC) recommend using 10% of the total blood volume as the maximum amount that can be collected at one time. To allow for an adequate recovery period, we further recommend that the maximum amount be withdrawn only once every 2 weeks (based on 10%). Daily samples can be collected as long as the cumulative blood volume does not exceed the calculated maximum over a 7-day period.

Place a tourniquet as high up on the rear leg of the mouse as possible. Apply a drop of artificial tears (petroleum jelly) on the lateral side of the leg and rub across the surface. The ointment will allow the blood to bead up on the skin. Using a 27-gauge needle, gently poke into the vein, taking care not to pierce completely through it because this will result in blood accumulating under the skin. As the blood beads up, you can use a heparinized capillary tube to collect the blood. When finished, release the tourniquet and apply pressure with gauze to stop the bleeding.

recommended supplies

- Heparinized capillary tube
- Clippers
- Artificial tears (petroleum jelly)
- Tourniquet
- 27-gauge needle
- Gauze

Voice-Over 1: Begin by clipping the lateral surface of the mouse's hind leg. Place a tourniquet as high up on the mouse's rear leg as possible. Apply petroleum jelly on the lateral side of the leg and rub across the surface as shown. The ointment will allow the blood to bead up on the skin.

Voice-Over 2: Identify the vein and with the needle make a solid, meaningful poke into the vein, being careful not to pierce completely through it as this will result in blood accumulating under the skin. As the blood beads up, you can use a heparinized capillary tube to collect the blood.

Voice-Over 3: When finished, apply pressure with gauze and release the tourniquet.

tail vein injection

Tail veins are one of the few intravenous injection sites on the mouse. Proficiency in one-handed injections is necessary; generally, small amounts of solution (<0.2 ml) are injected using this method.

The veins are located superficially on the lateral sides of the tail. To dilate the tail veins, begin by warming the tail with a warm gel pack or recirculating water blanket. Use of isoflurane anesthesia will also cause the veins to dilate. Start one-third from the distal end of the tail and work toward the base in the event that the initial try is unsuccessful. Insert the needle bevel side up into the vein while drawing back on the plunger. You will get a flash of blood into the needle hub when you are in the vein. Begin injecting; if the blood proximal to the needle vacates the vein, continue to inject. There will be virtually no resistance when correctly injecting into the vein. If a whitish bleb or bubble appears under the skin, stop injecting, remove the needle, and choose a site closer to the base of the tail.

recommended supplies

- 1- or 3-ml syringe
- 27- to 30-gauge needles
- Restraint tube
- Gel pack

Voice-Over 1: Begin by warming a gel pack in a microwave for approximately 30 seconds. Touch the pack to your wrist to ensure it is not too hot before wrapping it around the mouse's tail; warming the tail will cause the tail veins to dilate.

Voice-Over 2: Start one-third from the distal end of the tail and work toward the base in the event that the initial try is unsuccessful. Insert the needle bevel side up into the vein while drawing back on the plunger. You will get a flash of blood into the needle hub when you are in the vein. Begin injecting; if the blood proximal to the needle vacates the vein, continue to slowly inject. There will be virtually no resistance when correctly injecting into the vein. If a whitish bleb or bubble appears under the skin, stop injecting, remove the needle, and choose a site closer to the base of the tail.

retro-orbital bleed (anesthesia required)

The retro-orbital bleed is a commonly used method to collect a relatively large volume of blood from the mouse. See the Appendix on page 87 for information on calculating the blood volume and maximum blood sample volume of the mouse. The trainers at the Research Animal Resources Center (RARC) recommend using 10% of the total blood volume as the maximum amount that can be collected at one time. To allow for an adequate recovery period, we further recommend that the maximum amount be withdrawn only once every 2 weeks (based on 10%). Daily samples can be collected as long as the cumulative blood volume does not exceed the calculated maximum over a 7-day period. Anesthesia is required to perform this technique.

The thumb and forefinger of the operator's nondominant hand are used to pull the facial skin taut and cause the eye to protrude slightly. Breathing and color are monitored throughout the procedure to ensure that the airway is not compromised. Using the dominant hand, gently slide the tip of the capillary tube behind the eye at approximately a 30–45° angle. When the tip of the capillary tube contacts the bony floor of the orbit, gently twist the tip between the thumb and forefinger to rupture the orbital sinus. Blood will flow by capillary action into the tube. At the conclusion of the blood withdrawal, release the tension on the facial skin. Close the lid over the eye and gently apply a gauze pad until the bleeding has stopped. Reconfirm normal color and respiration and return the animal to its cage for recovery. Use alternate eyes for successive bleeds.

recommended supplies

- Heparinized capillary tube
- Tube rack
- Gauze
- Blood collection tube
- Anesthesia

Voice-Over 1: This illustration depicts the orbital sinus of the mouse. Rupturing these vessels shown in blue initiates the flow of blood.

Voice-Over 2: While holding the anesthetized mouse by the scruff, pull the facial skin taut, causing the eye to protrude. Gently insert the capillary tube behind the eye. When the tube contacts the bony floor of the orbit, twist until the orbital sinus is ruptured and blood flows through the tube and into the microcentrifuge tube. When finished, use gauze to close the eye, apply slight pressure, and stop the bleeding.

Voice-Over 3: Pull the facial skin taut, causing the eye to protrude. Slide the tip of the capillary tube behind the eye and, when the tube contacts the bony floor, gently rotate between the thumb and forefinger to rupture the orbital sinus. Blood flows by capillary action into the tube. When finished, use gauze to close the eye while applying slight pressure to stop the bleeding.

retro-orbital injections (anesthesia required)

The retro-orbital injection technique using anesthesia can be an alternative to tail vein injections. (Not all types of solutions may be injected.) Contact your lab animal veterinarian for guidelines in regard to the type of solutions you can inject. Proficiency in one-handed injections is necessary, and generally small amounts of solution are injected using this method (up to 200 µl, based on an adult mouse).

The thumb and forefinger of the operator's nondominant hand are used to pull the facial skin taut and cause the eye to protrude slightly. Breathing and color are monitored throughout the procedure to ensure that the airway is not compromised. Using a 1-ml syringe and a 27- or 30-gauge needle, with the bevel facing away from the eye, insert the needle at a 30–45° angle. Gently insert the tip of the needle until it contacts the bony floor of the orbit and slowly inject. Unlike most injections, it is not necessary to draw back on the plunger. Remove the needle carefully, keeping the bevel outward to protect the eye of the mouse from being scratched.

recommended supplies

- 1-ml syringe
- 27- or 30-g needle
- Anesthesia

Voice-Over: With the anesthetized mouse on the table, the thumb and forefinger of the operator's nondominant hand is used to pull the facial skin taut, causing the eye to protrude. Insert the needle, with the bevel facing away from the eye, until it contacts the bony floor of the orbit and slowly inject. It is not necessary to draw back on the plunger. Remove the needle carefully to avoid scratching the mouse's eye. In a matter of seconds, the blue dye that was injected in this demonstration can be seen in the feet, ears, and tail.

ear notching

Ear notching or punching is a long-term identification method that can be performed quickly with little pain or distress. Mice are restrained by the scruff, and using an ear punch, holes or notches are placed in the ears following a simple identification chart. Removed tissue can be used for genotyping, possibly replacing the need for a tail biopsy. Care must be taken when placing the punches. Punches too close to the edge of the ear may be ripped or torn open by the mouse, leaving patterns hard to identify.

recommended supplies

- Scissors-style or thumb-style ear punch

Voice-Over 1: [Pic. #1] This illustration shows an identification code that can be used to identify mice with numbers 1 through 399.

Voice-Over 2: [Pic. #2] Shown from top to bottom: ear tag applicator, ear tags, thumb- and scissor-style ear notchers.

Voice-Over 3: Punching or notching holes at various positions in the ears requires the use of an ear punch. Grasp the mouse securely by the scruff and, while following the code, place the notches or punches into the appropriate ear. In this demonstration, the mouse has been identified as 15. Check the markings regularly to ensure the mice can be identified accurately, especially if mice are group housed. During fights, the mice could rip each others' ears, making identification impossible. Removed tissue can be used for genotyping, possibly replacing the need for a tail biopsy.

ear tagging

Ear tagging is a long-term identification method that can be performed quickly with little pain or distress. Mice are restrained, and using an ear tag applicator, a uniquely numbered tag is placed in the lower one-third of the ear. Care must be taken when placing the tags. Tags could be torn out if placed too close to the edge of the ear.

recommended supplies

- Numbered tags
- Tag applicator

Voice-Over 1: This clip shows how to correctly place an ear tag into the applicator tip, followed by what the tag looks like after it has been crimped.

Voice-Over 2: After securing the mouse with the appropriate restraint, position the applicator over the lower one-third of the mouse ear and crimp. The ideal position will have the number facing forward for easy identification while the mouse is in the cage.

II

mouse procedural technique handouts

intraperitoneal (IP) injection

Purpose (what you will learn): How to safely administer an intraperitoneal (IP) injection.

Use this technique if: Using large volume of liquids (20 ml/kg), for example, if you are administering drug doses that have been diluted. Injection volume information can be found at http://www.rarc.wisc.edu.

Recommended skills: Basic handling and restraint of a mouse. Observe caution during the holding and injection process. Damage to internal organs can occur if care is not taken to restrain the animal and control the insertion of the needle.

Recommended supplies: A 25-gauge needle, 1-ml syringe, exam gloves, and sharps container.

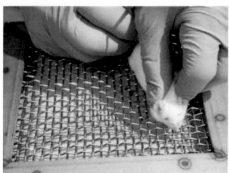

You must be able to adequately restrain a mouse before performing any injection technique.

By turning the mouse over, you can view the abdomen. Make sure the chest of the mouse is moving up and down and that you are not restricting breathing.

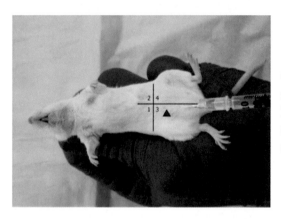

Quadrant 3 will be the best area for the IP injection. Inject halfway between the midline and where the leg attaches to the body.

The needle should be at a 45° angle. You will sense the needle pop through the skin and abdominal wall. Draw back slightly on the plunger. When you observe a vacuum bubble, then inject your solution.

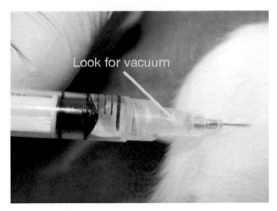

If while drawing back you get a colored liquid, pull the needle out, discard the needle and syringe, and start over with a new solution. Always dispose of the needle and syringe into a sharps container.

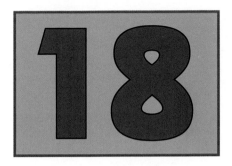

subcutaneous (SQ) injection

Purpose (what you will learn): How to safely administer a subcutaneous (SQ) injection.

Use this technique if: Injecting up to 10 ml/kg (maximum is 40 ml/kg) of liquid. Use several sites over the back of the animal if larger volumes must be administered, for example, if you are administering drug doses that have been diluted. Injection volume information can be found at http://www.rarc.wisc.edu.

Recommended skills: Basic handling and restraint. You must know and understand the proper methods to restrain a mouse for this technique.

Recommended supplies: A 25-g needle, 1-ml syringe, exam gloves, and sharps container.

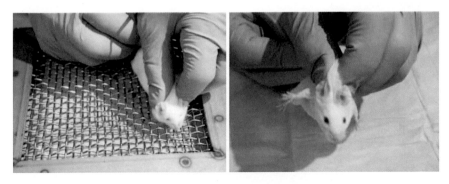

You must be able to adequately restrain a mouse before performing any injection technique.

Subcutaneous injections can be easily administered to mice. Restraining by the scruff is one technique. Grasp the scruff and insert the needle between the tented folds of skin. Pull back on the plunger to verify that a vacuum is created, then inject the solution.

If injections need to be placed in the back or above the tail area, the use of a towel is recommended. Place the mouse at the center of the towel and fold it over the mouse, leaving the tail area exposed.

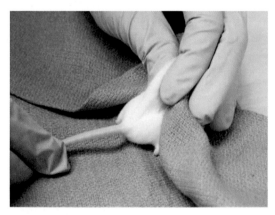

Restrain the front two-thirds of the mouse with the palm of your nondominant hand while tenting the skin over the hindquarters with your thumb and index finger.

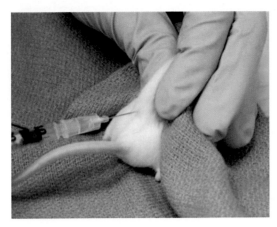

Insert the needle at the base of the tented folds of skin. Pull back on the plunger of the syringe to verify that a vacuum is created, then inject the solution.

maxillary bleed

Purpose (what you will learn): How to safely obtain a blood sample from the maxillary vein of the mouse.

Use this technique if: Collecting larger volumes of blood. See the Appendix on page 87 for information on calculating the blood volume and maximum blood sample volume of the mouse. The trainers at the Research Animal Resources Center (RARC) recommend using 10% of the total blood volume as the maximum amount that can be collected at one time. To allow for an adequate recovery period, we further recommend that the maximum amount be withdrawn only once every 2 weeks (based on 10%). Daily samples can be collected as long as the cumulative blood volume does not exceed the calculated maximum over a 7-day period.

Recommended skills: You must know and understand the proper methods to restrain a mouse for this technique. Multiple blood draws are possible from the same site as long as the maximum amount of total blood drawn is not exceeded.

Recommended supplies: A 5.5-mm lancet, blood collection tube, 2 × 2 gauze, exam gloves, sharps container.

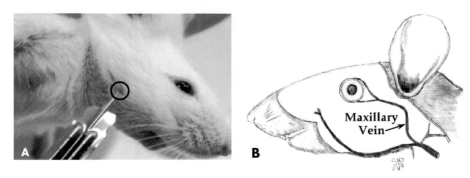

Albino mice have a convenient landmark shown in (A) that is in the area of the maxillary vein. The circle shows where to insert the lancet. Dark-colored mice have this landmark as well, but it may be harder to see. (B) shows the deep layer of the lateral surface of the mouse and identifies the location of the vein.

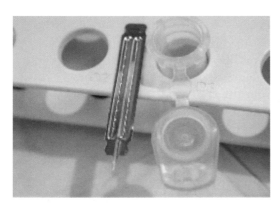

Minimum supplies are needed to perform the maxillary bleed: a lancet and a blood collection tube.

Proper restraint is one of the keys to success with this technique. As the skin around the neck is stretched, it applies pressure to the maxillary vein.

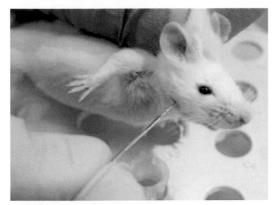

A meaningful poke is required to initiate the flow of blood. Maintaining the proper restraint aids the flow of blood.

The blood drops can be collected while holding the mouse over a blood collection tube. Approximately 3 drops equals 100 µl. Once the desired amount of blood is collected, release the mouse from your restraint and apply a clean compress to the bleeding site if necessary. Return the mouse to its cage.

oral gavage

Purpose (what you will learn): How to safely feed or dose by mouth using a feeding needle.

Use this technique if: You need to administer liquid food, experimental solutions, or medications directly to the stomach of a mouse. The recommended normal volume is 10 ml/kg. Dosing volume information can be found at http://www.rarc.wisc.edu.

Recommended skills: Basic handling and restraint of a mouse while inserting a gavage or feeding needle into the mouse esophagus.

Recommended supplies: A 22-gauge × 2-inch feeding needle (adult mouse), Luer lock syringe, exam gloves, and sharps container.

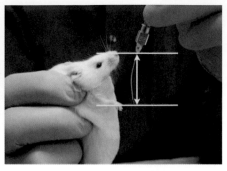

You must be able to adequately restrain a mouse before performing this technique.

Once the proper restraint is achieved, determine if the length of the needle is appropriate. The length of the needle needs to be approximately the same as the distance from the nose of the mouse to its stomach (just below the rib cage).

Left: Place the feeding needle in the back of the mouth and tilt the head back until the syringe is parallel with the body of the mouse. The needle will slide down easily when positioned properly. Do not use force. (It is possible to perforate the esophagus or force the needle into the trachea.) Proper insertion will be achieved using gravity and the weight of the syringe. If there is any resistance, stop, pull the needle out, and try again. Right: When the technique is done correctly, you will not see the gavage needle, only the hub, as shown in the picture. At this time, inject your solution. Slowly remove the needle, following its curve as you pull it out.

pedal vein blood draw

Purpose (what you will learn): How to safely obtain a blood sample from the pedal vein of the mouse. You will need a restraint device if an anesthesia machine is not used.

Use this technique if: Collecting relatively small volumes of blood (<50 μl). See the Appendix on page 87 for information on calculating the blood volume and maximum blood sample volume of the mouse. The trainers at the Research Animal Resources Center (RARC) recommend using 10% of the total blood volume as the maximum amount that can be collected at one time. To allow for an adequate recovery period, we further recommend that the maximum amount be withdrawn only once every 2 weeks (based on 10%). Daily samples can be collected as long as the cumulative blood volume does not exceed the calculated maximum over a 7-day period.

Recommended skills: Basic handling and restraint of a mouse. You must know how to properly use an anesthesia machine if the mouse will be anesthetized.

Recommended supplies: A 25-gauge needle, tourniquet, petroleum jelly or artificial tears, blood collection tube, heparinized capillary tube (inside diameter 1.1 mm, wall 0.2 × 75 mm L), exam gloves, anesthesia machine, gauze, and sharps container.

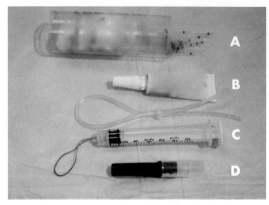

Supplies for the pedal vein blood draw:

A. Heparinized capillary tube

B. Artificial tears

C. Tourniquet

D. 25- or 27-gauge needle

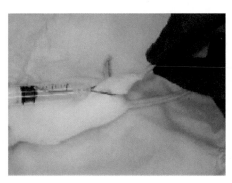 

Left: Place a tourniquet as high up on the leg of the mouse as possible. Right: Apply the artificial tears or petroleum jelly to the leg as shown. Rub the artificial tears around the top of the foot. The ointment will allow the blood to bead up on the skin surface instead of spreading across the foot.

Moving in the direction of the arrow, gently poke into the vein, taking care not to pierce completely through because this will cause blood to collect under the skin.

As the blood beads up on the foot, use the heparinized capillary tube to collect the blood.

When finished, release the tourniquet and wipe with the gauze. Apply momentary pressure to stop the bleeding.

saphenous vein blood draw

Purpose (what you will learn): How to safely obtain a blood sample from the saphenous vein of the mouse. You will need a restraint device if an anesthesia machine is not used.

Use this technique if: Collecting relatively small volumes of blood (<50 µl). See the Appendix on page 87 for information on calculating the blood volume and maximum blood sample volume of the mouse. The trainers at the Research Animal Resources Center (RARC) recommend using 10% of the total blood volume as the maximum amount that can be collected at one time. To allow for an adequate recovery period, we further recommend that the maximum amount be withdrawn only once every 2 weeks (based on 10%). Daily samples can be collected as long as the cumulative blood volume does not exceed the calculated maximum over a 7-day period.

Recommended skills: Basic handling and restraint of a mouse. You must know how to properly use an anesthesia machine if the mouse will be anesthetized.

Recommended supplies: A 27-gauge needle, tourniquet, petroleum jelly or artificial tears, blood collection tube, heparinized capillary tube (inside diameter 1.1 mm, wall 0.2 × 75 mm L), exam gloves, anesthesia machine, gauze, small clippers, and sharps container.

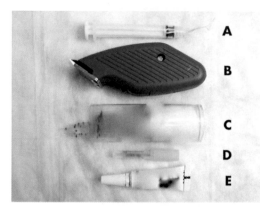

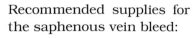

Recommended supplies for the saphenous vein bleed:

A. Tourniquet

B. Clippers

C. Heparinized capillary tube

D. 27-gauge needle

E. Artificial tears

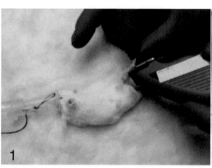

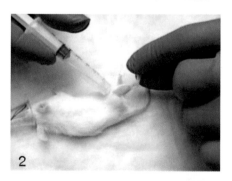

Clip the lateral surface of the hind leg of the mouse (1). Place a tourniquet as high up on the leg as possible (2). Apply the artificial tears to the leg as shown (3) and spread on the lateral surface. The ointment allows the blood to bead up on the skin surface instead of spreading across the leg.

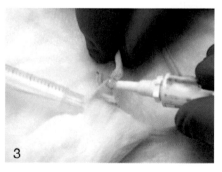

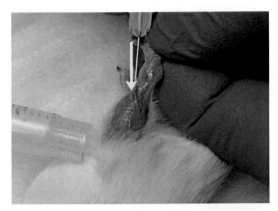

Moving in the direction of the arrow, gently poke into the vein, taking care not to pierce completely through because this will cause blood to collect under the skin.

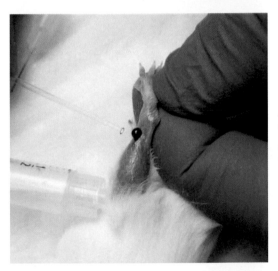

As the blood beads up on the leg, use the heparinized capillary tube to collect the blood.

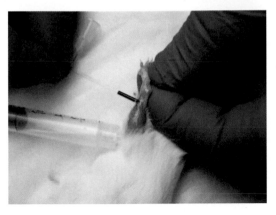

When finished, release the tourniquet and wipe with the gauze. Apply momentary pressure to stop the bleeding.

retro-orbital blood draw

Purpose (what you will learn): How to safely obtain a blood sample from the orbital plexus of the mouse.

Use this technique if: Collecting larger volumes of blood. See the Appendix on page 87 for information on calculating the blood volume and maximum blood sample volume of the mouse. The trainers at the Research Animal Resources Center (RARC) recommend using 10% of the total blood volume as the maximum amount that can be collected at one time. To allow for an adequate recovery period, we further recommend that the maximum amount be withdrawn only once every 2 weeks (based on 10%). Daily samples can be collected as long as the cumulative blood volume does not exceed the calculated maximum over a 7-day period.

Recommended skills: You must know and understand the proper methods to restrain a mouse for this technique and how to properly use an anesthesia machine.

Recommended supplies: A blood collection tube, heparinized capillary tube (inside diameter 1.1 mm, wall 0.2 × 75 mm L), 2 × 2 gauze, exam gloves, anesthesia machine, and sharps container.

Orbital Sinus

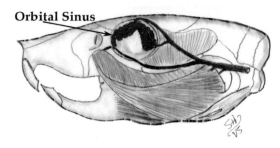

This drawing shows the orbital sinus that lies behind the eye of the mouse.

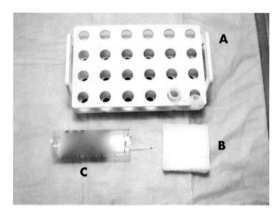

Recommended supplies for the retro-orbital bleed:

A. Blood collection tube

B. 2 × 2 gauze

C. Heparinized capillary tube

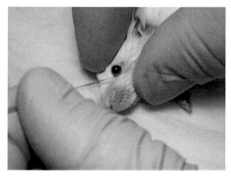
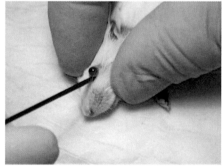

You can position the mouse in one of two ways. Images above show the mouse on a table; the skin around the eye is stretched to the sides, allowing the eye to bulge slightly. The capillary tube is advanced behind the eye, then gently twisted to break the vessels; the capillary tube fills with blood. You can transfer the blood into a blood collection tube or spin it down for plasma. Being overly aggressive can cause permanent eye damage.

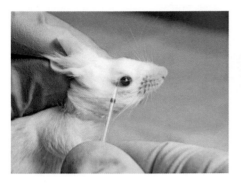 

The technique in the sequence of pictures above shows how the blood is allowed to flow down the capillary tube and drip directly into the blood collection tube. Use the capillary tube to displace the eye while advancing the tube toward the vessels.

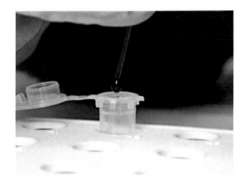

When finished, use the gauze and hold the eye closed to stop the bleeding. Allow the mouse to fully recover and monitor closely to ensure all bleeding has stopped.

retro-orbital injection

Purpose (what you will learn): How to safely inject solution into the orbital sinus of the mouse.

Use this technique if: You need to inject into the mouse intravenously. Can be used as an alternative to the tail vein injection. Proficiency in one-handed injections is necessary, and only small amounts of solution are injected using this method (up to 200 μl, based on an adult mouse).

Recommended skills: You must know and understand the proper methods to restrain a mouse for this technique. You must know how to properly use an anesthesia machine because it is a requirement that mice be anesthetized for this procedure.

Recommended supplies: A 27-gauge needle, 1-ml syringe, exam gloves, gauze, anesthesia machine, and sharps container.

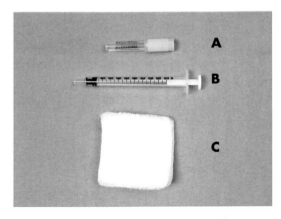

Recommended supplies for the retro-orbital injection:

A. 27-gauge needle

B. 1-ml syringe

C. 2 × 2 gauze

D. Anesthesia machine (not pictured)

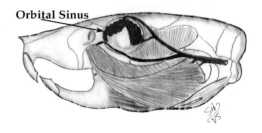

Orbital Sinus

This drawing shows the orbital sinus that lies behind the eye of the mouse. This is the area where the needle is inserted.

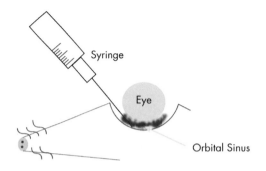

Syringe

Eye

Orbital Sinus

This illustration shows the placement of the needle into the cranial portion of the orbital sinus.

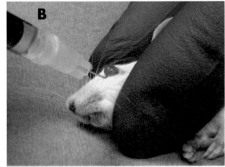

A

B

The thumb and forefinger of the operator's hand is used to pull the facial skin taut and cause the eyeball to protrude slightly (A). The respiratory rate and color of the mouse are monitored throughout the procedure to ensure that the airway is not compromised. Using a 1-ml syringe and a 27-gauge needle, with the bevel facing away from the eyeball, insert the needle at a 30–45° angle (B). Gently insert the tip of the needle until it contacts the bony floor of the orbit and slowly inject. Unlike most injections, it is not necessary to draw back on the plunger. Remove the needle carefully, keeping the bevel outward to protect the eyeball of the mouse from being scratched.

When finished, use the gauze and hold the eyelid closed for a few seconds. Monitor the mouse until it fully recovers, placing it back into the cage only after it is able to ambulate on its own.

tail vein injection

Purpose (what you will learn): How to safely inject solutions into the lateral tail vein of the mouse.

Use this technique if: You need to inject solutions into the mouse intravenously. The bolus volume is 5 ml/kg; the slow injection volume is 25 ml/kg.

Recommended skills: You must know and understand the proper methods to restrain a mouse for this technique. Experience using an anesthesia machine is helpful.

Recommended supplies: A 27-gauge needle, 1-ml syringe, gel pack or heated water blanket, centrifuge tube with a 3-mm hole cut in the end or a mouse-specific restraint, exam gloves, anesthesia machine may be used if in animal protocol, and sharps container.

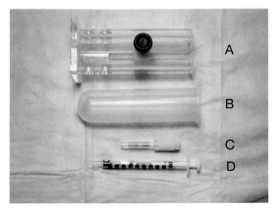

A
B
C
D

Supplies that are needed for the tail vein injection:

A. Acrylic mouse restraint

B. Centrifuge tube (for restraint)

C. 27-gauge needle

D. 1-ml syringe

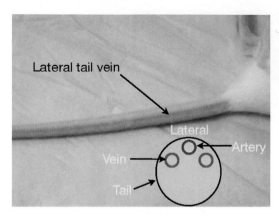

In this picture the lateral tail vein is obvious. Warm the tail using a gel pack, light source, or recirculating water blanket to dilate the blood vessels. If isoflurane gas anesthesia is used, it will also cause the veins to vasodilate.

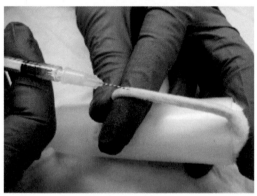

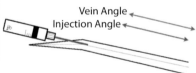

Insert the mouse head first into the restraint tube as shown in the picture; use your middle finger to keep the mouse from backing out. Start at the most distal end of the tail and work toward the base if the initial try is unsuccessful.

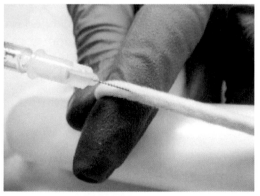

Insert the needle into the vein, draw back slightly on the plunger, and look for blood to flash back into the needle hub. Begin injection; if the blood proximal to the needle vacates the vein, continue to inject. There will be virtually no resistance when correctly injecting into the vein. If a bubble of solution appears under the skin, stop injecting, remove the needle, and choose a site closer to the base of the tail.

The tail vein injection, like the tail artery draw, can be frustrating for the beginner. Practice and use of the appropriate equipment will make the procedure easier. Be sure to take advantage of a heat source as mentioned as this will greatly increase your chance of identifying the vein.

ear notching/punching and ear tags

Purpose (what you will learn): How to safely perform ear notching/punching as well as placing ear tags in rodents.

Use this technique if: You need to individually identify mice or rats.

Recommended skills: You must know and understand the proper methods to restrain a mouse for this technique.

Recommended supplies: Thumb- or scissor-style ear punch or ear tag applicator and numbered ear tags.

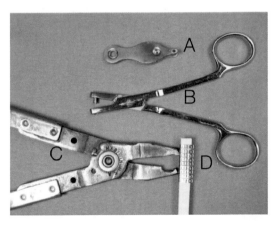

Recommended supplies for ear notching or ear tagging:

A. Thumb-style ear punch

B. Scissor-style ear punch

C. Ear tag applicator

D. Numbered ear tags

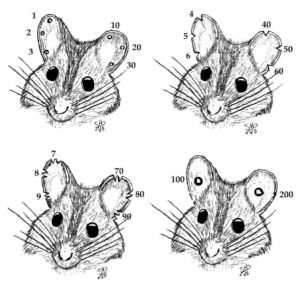

This illustration shows a rodent identification code that can be used to identify rodents with numbers 1 through 399.

ear notching/punching

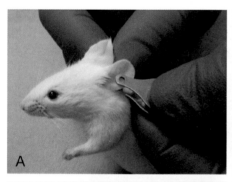

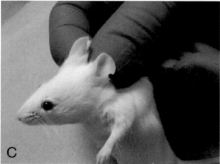

Ear notching/punching is a long-term identification method that can be performed quickly with little pain or distress. In this series of pictures, the mouse is restrained by the scruff (A), and using an ear punch, holes (B) or notches (C) are placed in the ears following the previously shown chart. Care must be taken when placing the punches.

Consistency in placing the notches/punches is necessary to allow for easy reading. Punches too close to the edge of the ear may be ripped or torn open, leaving patterns hard to identify.

ear tagging

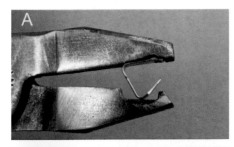

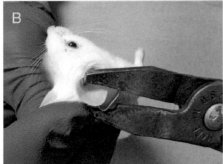

Ear tagging is also a long-term identification method that can be performed quickly with little pain or distress. The first picture in the series at left shows the proper placement of the identification tag in the applicator (A). The next two pictures demonstrate placing the tag in the lower one-third of the ear of the mouse (B) and then the tag secured in the ear (C). Place the tag on the lower portion of the ear to prevent the ear from folding over. Notice that the tag is positioned with the number facing forward, allowing it to be viewed easily. Care must be taken when placing the tags. Tags could be torn out if placed too close to the edge of the ear.

Note: Although this handout uses the mouse for demonstration purposes, the techniques used can be easily applied to rats. The only difference is in the manner of restraint.

appendix a: mouse (Mus musculus)

Genus and species: *Mus musculus*, Order: Rodentia

About 95% of research animals are rodents, with 90% being mice and rats.

Qualities that make the mouse a good model for research:

- Care is easy.
- Small size makes them easy to use.
- They have high reproductive capacity.
- They have a short reproduction time.
- An abundance of baseline information already exists.
- A large number of strains are available with natural genetic deficiencies that can act as models for humans.
- There are germ-free and pathogen-free production techniques.

Common inbred strains of mice:

- BALB/c: albino in color (white coat and pink eyes)
- C3H: agouti in color (dark brown with yellow bands close to tips of hair)
- C57Bl/6: black in color
- DBA: the oldest inbred strain; brown in color
- 129, FVB (transgenic)

Common outbred strains of mice:

- ICR (CD-1)
- Swiss Webster
- Nude
- SCID

Genotyping:

- Tail snip is done prior to 21 days old; no anesthesia is required. Older mice would have to be anesthetized.
- DNA samples can be acquired by amputating 5 mm or less of the distal tail.
- Tissue removed from ear punching can be used for polymerase chain reaction for rapid screening of mice.
- Hair, saliva, or feces can be used in place of tail snipping.
- The blade/scissors need to be sanitized between animals.

physical characteristics/anatomical features[1]

Some general features for mice are as follows:

- **Acute hearing:** Sensitive to ultrasounds and high-pitched noises
- **Smell**: Well-developed sense of smell
- **Different vision**
- **Open-rooted incisors:** Grow continually throughout life and have enamel only on the anterior surface, which results in continuous sharpening of the incisors as they grind against one another
- **Males** have open inguinal canal
- **Tail** is principal organ for thermoregulation
- **Cannot vomit**
- **Resting heart rate**: 300+ per minute and **respiratory rate** 100+ per minute

---

[1] See appendix b normative data for the laboratory mouse.

- **Weight**: Average adult mouse 20–40 g; male mice slightly larger than female mice
- **Life span**: On average 1–3 years
- **Teeth**: 16 teeth: 2 upper and 2 lower incisors, 3 upper and 3 lower molars on each side
- **Mammary glands**: 5 pair for female mice: 3 pair thoracic and 2 pair abdominal
- **Mammary gland tissue**: Extends over the back and shoulders of female; mammary gland tumors often on the side or along the back of the mouse
- **Spleen**: Male mouse spleen approximately 50% larger than that of the female mouse

behavior

- Mice are generally docile.
- **Nocturnal:** Most of their activity takes place at night.
- Female mice usually do not fight when housed together.
- Male mice will often fight when housed together. Sometimes adding environmental enrichment alleviates this behavior. Closely monitor these mice to be sure the enrichment is effective. Otherwise, separate aggressive mice as they could possibly inflict severe injuries on cage mates. Housing male mice together from weaning or birth may prevent fighting as they mature.
- **Barbering**: A behaviorally dominant mouse will sometimes chew the fur off a subordinate mouse. This is usually harmless. If the dominant mouse is removed from the cage, the hair on the subordinate mouse will grow back. Barbering should not be confused with fur loss caused by fighting, skin problems, or mites. Adding environmental enrichment may help prevent barbering.
- Mice are highly influenced by sound and pheromones (substances produced by the mouse and excreted in the urine). Mice use pheromones to attract or get a response from another mouse.
- Mice normally sit on all four feet with their eyes and ears alert.

- A mouse normally sleeps curled up with its head tucked under its body. If mice are group housed, they may sleep side by side or on top of one another.
- Mice have a high metabolic rate and are very active when they are awake. They are constantly burrowing, foraging, or moving things around their cage and are almost constantly grooming themselves and one another.

sexing and breeding

- **Anogenital distance** is used to determine male and female. The distance between the anus and genital papilla is twice the distance in the male than in the female.
- **Females have visible mammary glands**; males' mammary glands are not visible.
- **Two** mating systems are commonly used.
  - **Monogamous**: One female is bred with one male.
  - **Polygamous** (harem breeding): Two or more females are bred with one male.
- **Sexual maturity**: Sexual maturity is reached at 40–60 days of age.
- Females are **polyestrous**; they come into heat every 4–5 days.
- **Gestation**: The gestation period is 19–21 days.
- Breeding mice usually build nests if bedding, such as nestlets, are available.
- Provide breeder chow for the pregnant and nursing females.
- Keep in mind that light cycles can affect breeding.
- **Vaginal plugs** are a good indication of a successful breeding.
- Shortly after pups are born, a distinct white spot should be visible through their transparent skin on the left side of the abdomen. This is called a milk spot and shows that their stomach is full of milk, which can be a good indication that the pups are healthy and nursing.
- If the mother is stressed or the pups are unhealthy, she may eat one or all of her pups (cannibalism). Try not to disturb the

mother or her newborn litter for several days after she gives birth.

- Females will go through a **postpartum estrus** within 24 hours of giving birth.

- If the female is bred during her postpartum estrus, she will usually deliver her next litter when it is time to wean her first litter.[2]

- Pups generally should be **weaned** at 21–28 days.

- Breeding life can be 12–18 months; litter sizes may decrease as the mouse gets older.

- **Litter size** can range from 4 to 12 but can be more or less, depending on the strain; transgenics, knockouts, and inbred strains tend to have litters on the lower end of the range.

- Newborn pups are pink, hairless, and helpless. Their eyelids and ear canals are sealed, and the external part of their ears is not fully developed.

- **Pup hair** begins to appear between 2 and 3 days of age.

- **Pup ears** open between 3 and 4 days of age.

- **Pup eyes** open between 12 and 14 days of age.

sources

Some of the more common commercial vendors are

Charles River Laboratories
Wilmington, MA 01887
1-800-522-7287
http://www.criver.com
Established strains, transgenics

Harlan Sprague Dawley/Harlan Teklad
Indianapolis, IN 46229, Madison, WI 53744
1-800-793-7287, 1-608-277-2070
http://www.harlan.com
Established strains, transgenics, diets

Jackson Laboratories
Bar Harbor, ME 04609
1-800-422-6423
http://jaxmice.jax.org/index.html
Established strains, mutant stocks

Taconic Farms
Germantown, NY 12526
1-888-822-6642
http://www.taconic.com
Established strains, transgenics

Noncommercial options are[3]:

- Other universities

- Other researchers on campus

---

[2] Always wean her older pups before she gives birth to her next litter. This will prevent overcrowding and allow the mother to give her full attention to the new litter.

[3] Make sure to discuss with your veterinarian before obtaining animals.

husbandry

Work quietly around the mice; they are very sensitive to sharp, loud noises.

There are various types of caging:

Shoebox cages: Most mice are housed in this type.

Wire-bottom cages: These are not allowed unless it is a research necessity to collect feces or urine or to prevent contact with the bedding.

Microisolator cages: These are used to house the mice to prevent disease transmission. This is especially important for immunocompromised animals. Mice are also provided with sterile food and water.

- When moving cages, it is important to remove or turn the water bottle around to prevent spilling water into the cage.
- If cages have automatic watering systems, make sure that the cage is pushed back all the way onto the rack to ensure that the water valve reaches into the cage and is accessible to the animal.
- Make sure that the water valve is not plugged or leaking.
- An assortment of commercial bedding is available.
- Cage changes should be done as described in the departmental standard operating procedures (SOPs).
- Rooms should be cleaned and disinfected according to facility SOPs.

diet

- Generally, mice are fed and watered ad libitum (continuous supply of food and water).
- Diets are normally provided in 4- to 5-g pellets. Normal adult **food intake** is 3–6 g/day. Pellets are firm and require gnawing, which helps keep the incisors worn down.
- Nonpelleted meals or powder diets are generally used when food intake is being monitored or when experimental substances

are being added to the diet. Remember to offer something for the mouse to chew or gnaw if you provide these diets.

- Water can be provided by water bottles or by an automatic watering system.
- Laboratory animal feed manufacturers produce a variety of nutritionally complete diets. Supplementing these diets is usually unnecessary.
- Depending on the cage design, food pellets can be placed in the feeders or in the designated area in the wire cage top. They should be filled with enough food to last several days. It is not necessary to fill food pellets to the top of the wire top; this can be wasteful.

handling and restraint

Due to a large quantity of loose skin on the neck, the mouse can be a challenge to restrain.

With that in mind, a mouse can be picked up by

- The base of the tail with your thumb and forefinger
- Grasping the base of the tail between the ends of a smooth-tipped forceps
- Grasping the skin of the mouse at the back of the neck with fingers or forceps
- Cupping your hands and picking up the group (along with some bedding) or grasping skin across the shoulder blades with forceps if pups are involved

identification

- **Cage cards** should be easy to read and identify. They should include sex, strain, identification number, investigator, protocol number, and so on.
- **Cage alerts** can provide an alert to special needs (e.g., if sick, pregnant, weaning, etc.).
- **Tape on cages is discouraged.**

- **Ear notch**: Animals do not need to be anesthetized; removed tissue can be used for polymerase chain reaction. See the rodent identification code in Chapter 26.
- **Ear tag**: Animals do not need to be anesthetized.
- **Tattoo**: Animals must be anesthetized.

stress management and enrichment

- Add enrichment such as nylon bones, ping-pong balls, or PVC pipe.
- Use nestlets in cages, not cotton.
- Try to service the room at the same time each day.
- Be quiet in their environment.

recognizing pain and distress

Most animals will try not to show pain or illness until they are quite sick. Most prey animals turn into victims if they show signs of weakness. The following are some common signs of pain or distress:

- **Abnormal biting**: Does the animal chew or self-mutilate?
- **Posture**: Is the animal hunched over or arching its back? Does it appear to be walking on "eggshells"?
- **Anxious**: Does the animal seem agitated and restless?
- **Aggressive**: Does the animal try to bite when touched?
- **Depressed**: Does the animal seem lethargic or that it does not care what you do to it?
- **Change in grooming**: Does the animal lack normal grooming behavior? Does the hair coat of the animal appear dull, ungroomed, and oily? Are there signs of hair loss?
- **Weight loss/dehydration**: Is there a decrease in food or water consumption?
- **Mammary tumors**: Are mammary tumors present?
- **Malocclusion**: Do the teeth of the animal line up, or are they growing in all different directions?
- **Bite wounds**: Are there bite wounds?

- **Parasites**: Are there parasite bites?
- **Excretions**: Does there appear to be an abnormal amount of feces or urine?
- **Coordination**: Is the animal unsteady or wobbly?
- **Movement**: Is the animal difficult to rouse, or does it seem restless?

common diseases and prevention

Diseases

- **EDIM** (epizootic diarrhea of infant mice virus)
- **MHV** (mouse hepatitis virus)
- **MPV** (mouse parvovirus)
- **MVM** (minute virus of mice)
- **Sendai**
- **Hantavirus**

Prevention

- **Use microisolator lids** on cages: only open the cages in a hood
- **Sentinel program**
- **Limit exposure**

record keeping

Records should be kept for all species; they should be available within close proximity to the animals and easily accessible to all lab personnel.

- Keep accurate, up-to-date records of everything that happens in a laboratory setting. This information helps facility managers and investigators determine whether procedures are followed according to established standards.
- **Husbandry records**, such as an **animal room log sheet**, can be used to identify the animal strains or species kept in

the room, where the animal came from, animal census, room temperature and humidity, cage changes, food and water changes, or when racks were washed, floors disinfected, or any other husbandry tasks accomplished.

- **Standard operating procedures (SOPs)** are documents that state how a procedure is to be performed. These procedures are written to allow different people to perform the same tasks in the same way at any time. SOPs should be kept in a central location where technicians and investigators have easy access to them.

- **Clinical records (experimental and surgical records)** should be maintained on all species. These records should indicate species, sex, and date of birth, health status, vaccinations, surgical procedures, postoperative care, and any other pertinent information.

- The National Institutes of Health (NIH) requires that all assurance records related directly to grant applications, research proposals, and changes of research activities be maintained for at least 3 years after completion of an activity.

- If an inspector asks for your records and you cannot provide a written copy regarding what you have done, the inspector must assume that it was not done. **If it is not written down, it never happened**.

---

protocols

Every research lab associated with a university must have an approved animal care and use protocol to perform research on live animals. With that in mind,

Read your protocol. The protocol contains the detailed information in regard to the approved techniques and procedures you will be performing.

euthanasia

You must follow the approved euthanasia regimen that is listed in your protocol. The following are **common euthanasia methods**:

- Inhalant anesthetics
- CO, CO_2
- Barbiturate overdose
- Cervical dislocation (mouse must be anesthetized)
- Decapitation

There are other acceptable methods of euthanasia that may be more useful to your research. Contact your veterinarian for suggestions or go to the following link for information from the American Veterinary Medical Association (AVMA) Panel of Euthanasia: http://www.avma.org/resources/euthanasia.pdf.

appendix b: normative data for the laboratory mouse

general information

Adult weight
   Male   20–40 g
   Female   18–35 g
Surface area   Weight $(g)^{2/3} \times (9) \times 10^4$
Body temperature   36.5–37.7°C
Food consumption   15 g/100 g/day
Water consumption 15 ml/100 g/day

blood and oxygen

Pressure
   Systolic   133–160 mm Hg
   Diastolic   102–110 mm Hg
Volume
   Plasma   3.15 ml/100 g body weight
   Whole blood   5.85 ml/100 g body weight
Heart rate   310–840 beats per minute
Tidal volume   0.18 (0.09–0.38) ml
Minute volume   24 (11–36) ml/minute
Stroke volume   1.3–2.0 µl/beat
Plasma
   pH   7.2–7.4
   CO_2 21.9 mmol/l
Respiration rate   163 (60–220) beats per minute
O_2 consumption   1.69 ml O_2/gm/hour

experimental information

Maximum single bleed[1]	10% total blood volume
Gavage volume	10 ml/kg

hematology

Leukocyte count	$8.4 \ (5.1–11.6) \times 10^3/\mu l$
Total	
Neutrophils	17.9% (6.7–37.2%)
Lymphocytes	69% (63–75%)
Monocytes	1.2% (0.7–2.6%)
Red blood cells	$8.7–10.5 \times 10^6/mm^3$
Hemoglobin	13.4 (12.2–16.2) g/dl
Platelets	$600 \ (100–1,000) \times 10^3/\mu l$
Packed cell volume	44% (42–44)

breeding

Breeding onset	50–60 days
Pseudopregnancy	10–13 days
Fertilization	2 hours postmating
Implantation	4–5 days
Gestation	19–21 days
Postpartum estrus	24 hours
Puberty	
Male	28–49 days
Female	28–49 days
Breeding life	12–18 months
Litter size	4–12
Birth weight	1–1.5 g
Eyes open	12–14 days
Wean	21 days

Recommended blood techniques for the laboratory mouse include use of the maxillary, saphenous, and pedal veins; tail vein/artery; and retro-orbital, jugular, and cardiac puncture.

For further information, please contact your lab animal veterinarian.

---

[1] Example for a 25-g mouse: Total blood volume (TBV) is 72 ml/kg × 0.025 kg = 1.8 ml.

appendix c: blood volume

Although additional percentages are given in this appendix, the trainers at the Research Animal Resources Center (RARC) recommend using 10% of the total blood volume as the maximum amount that can be collected at one time. To allow for an adequate recovery period, we further recommend that the maximum amount be withdrawn only once every 2 weeks (based on 10%). Daily samples can be collected as long as the cumulative blood volume does not exceed the calculated maximum over a 7-day period.

Circulating blood volume in the mouse:[1]

Species	Blood Volume (ml/kg)	
	Recommended Mean	Range of Mean
Mouse	72	63–80

Blood volume based on a 25-g mouse:
72 ml/kg × 0.025 kg = 1.8 ml

Total blood volume and recommended maximum blood sample volume:[1]

Species	Blood Volume (ml)	7.5% (ml)	10% (ml)	15% (ml)
Mouse (25 g)	1.8	0.1	0.2	0.3

---

[1] EFPIA/ECVAM paper on good practice in administration of substances and removal of blood. *J Appl Toxicol* 21: 15–23, 2001.

Mouse blood collection guidelines by percentage of total blood volume:[2]

weekly	7.5%
every 2 weeks	10%
every 4 weeks	15%

[2] American Association for Laboratory Animal Science. *Laboratory Mouse Handbook.* Memphis, TN: American Association for Laboratory Animal Science, 2006.